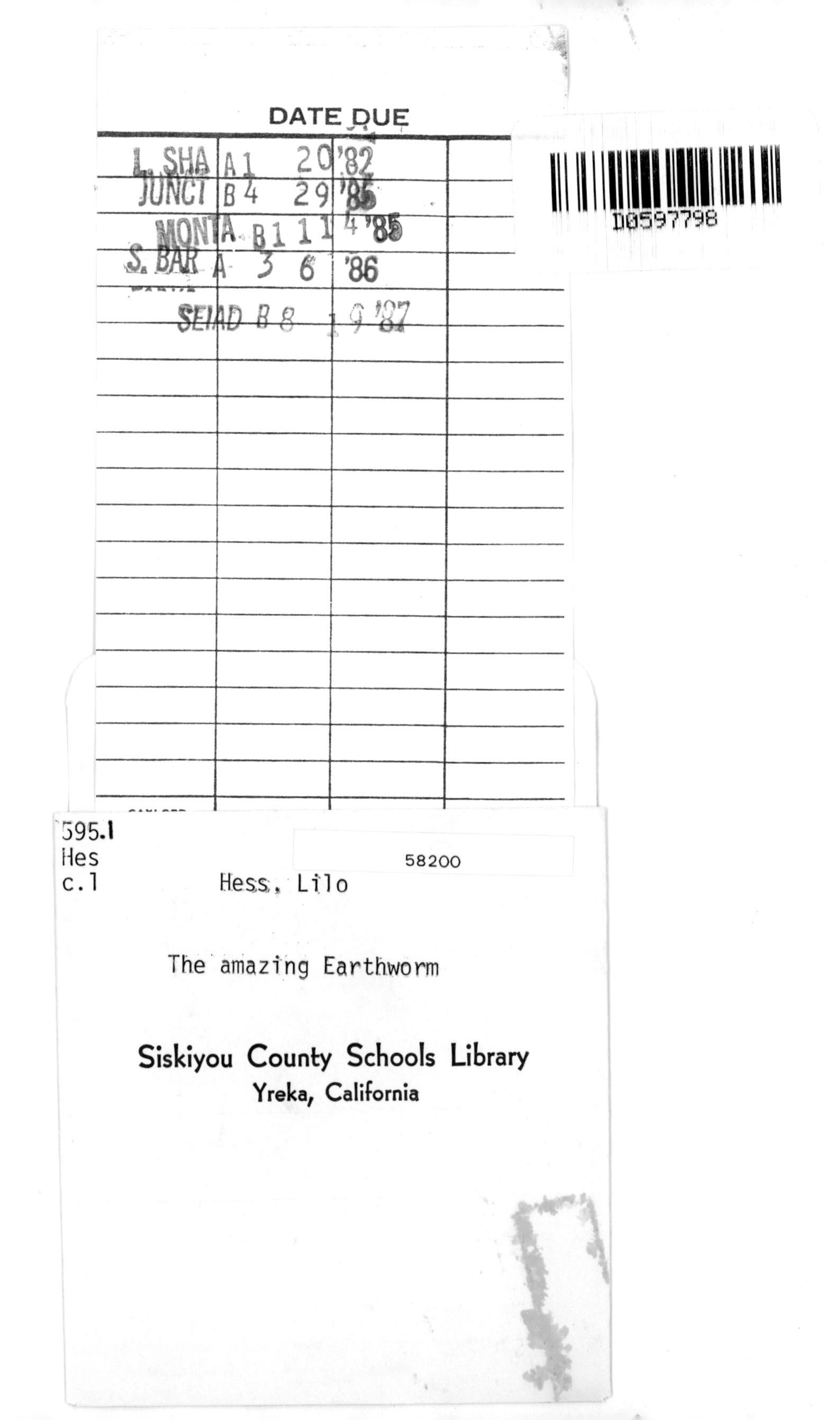
DATE DUE
L. SHA A1 20'82
JUNCT B4 29'85
MONTA B11 4'85
S. BAR A 3 6 '86
SEIAD B8 19'87
D0597798

The Amazing Earthworm

The Amazing Earthworm

LILO HESS

Charles Scribner's Sons, New York

Library of Congress Cataloging in Publication Data

Hess, Lilo.
The amazing earthworm.

SUMMARY: Describes the characteristics and habits of that best-known worm, the earthworm.
1. Earthworms—Juvenile literature [1. Earthworms]
I. Title.
QL391.A6H47 595'.146 78-24321
ISBN 0-684-16079-X

This book published simultaneously in the United States of America and in Canada—
Copyright under the Berne Convention

1 3 5 7 9 11 13 15 17 19 MD/C 20 18 16 14 12 10 8 6 4 2

Printed in the United States of America

The Amazing Earthworm

AUSTRALIAN NEWS AND INFORMATION BUREAU

Worms are simple animals, but they are very numerous and varied. There are roundworms, flatworms, hairworms, hookworms, ribbon worms, fanworms, ringed worms, and many others. The ringed or segmented worms alone have about 7000 varieties.

Some worms live on land, some in the ocean or in freshwater ponds and streams. Some burrow into the soil, others live under stones or in rock cavities. Some live on or inside animals, even humans. Some are so small that you can hardly see them; others, like the giant Australian earthworm, may grow to be 12 feet (3.6 meters) long.

One of the best known and probably the most useful of the worms is the earthworm. The earthworm is a ringed or segmented worm and belongs to the group called Annelids, meaning "ringed." The worm's rings or segments are clearly visible. They help the worm to twist, and wiggle forward or backward.

Neither earthworms nor other worms have backbones. They belong to a group which is called the invertebrates. Jellyfish, snails, and oysters are also invertebrates.

The earthworm is long and round, and at first it is not easy to distinguish which end forms the head and which end forms the tail. If you stand quietly and watch an earthworm move, you can see that the slightly pointed end goes forward first. This is the head. The worm also has a top side and a belly side. If you turn a worm over, it will immediately right itself.

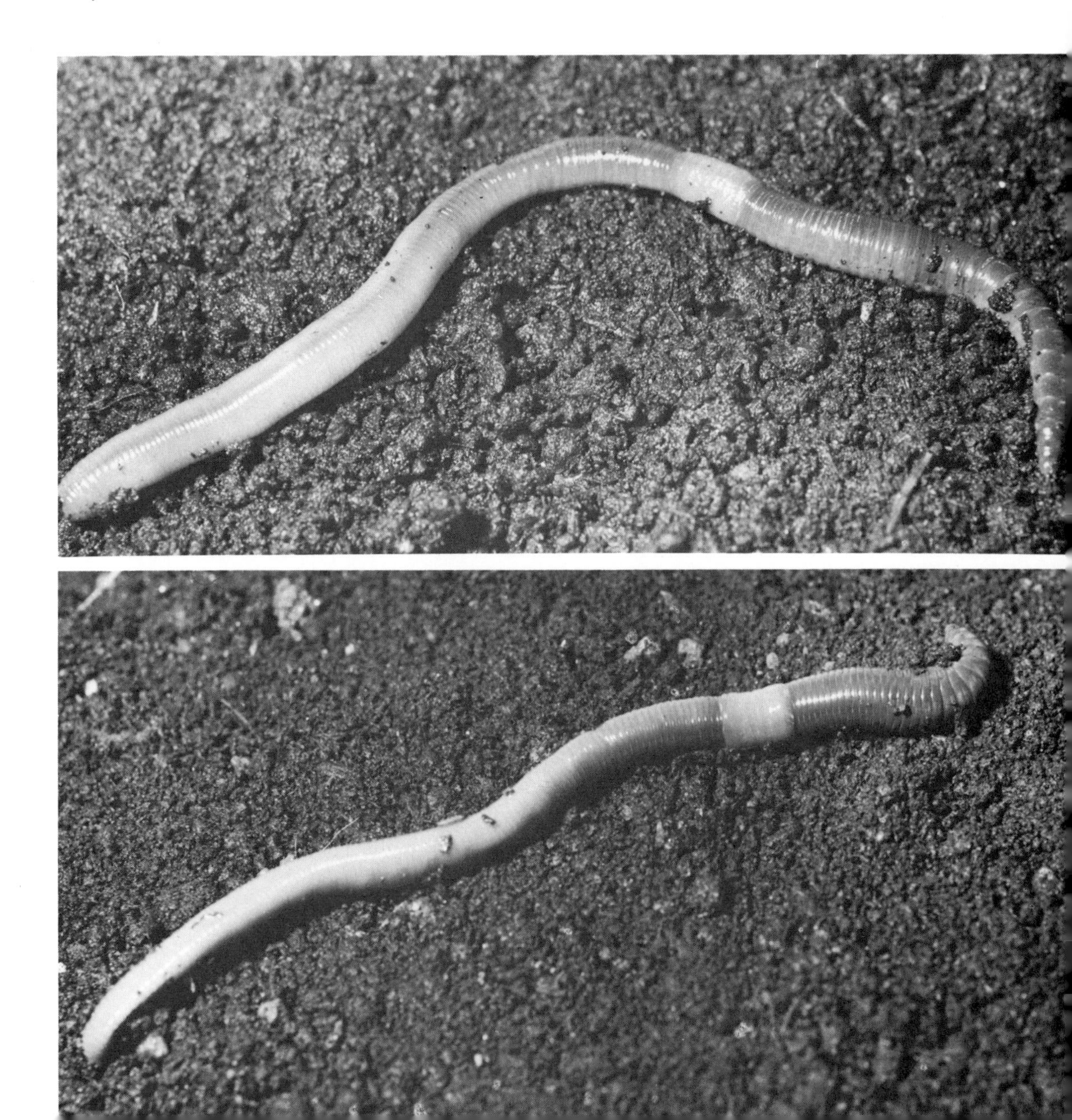

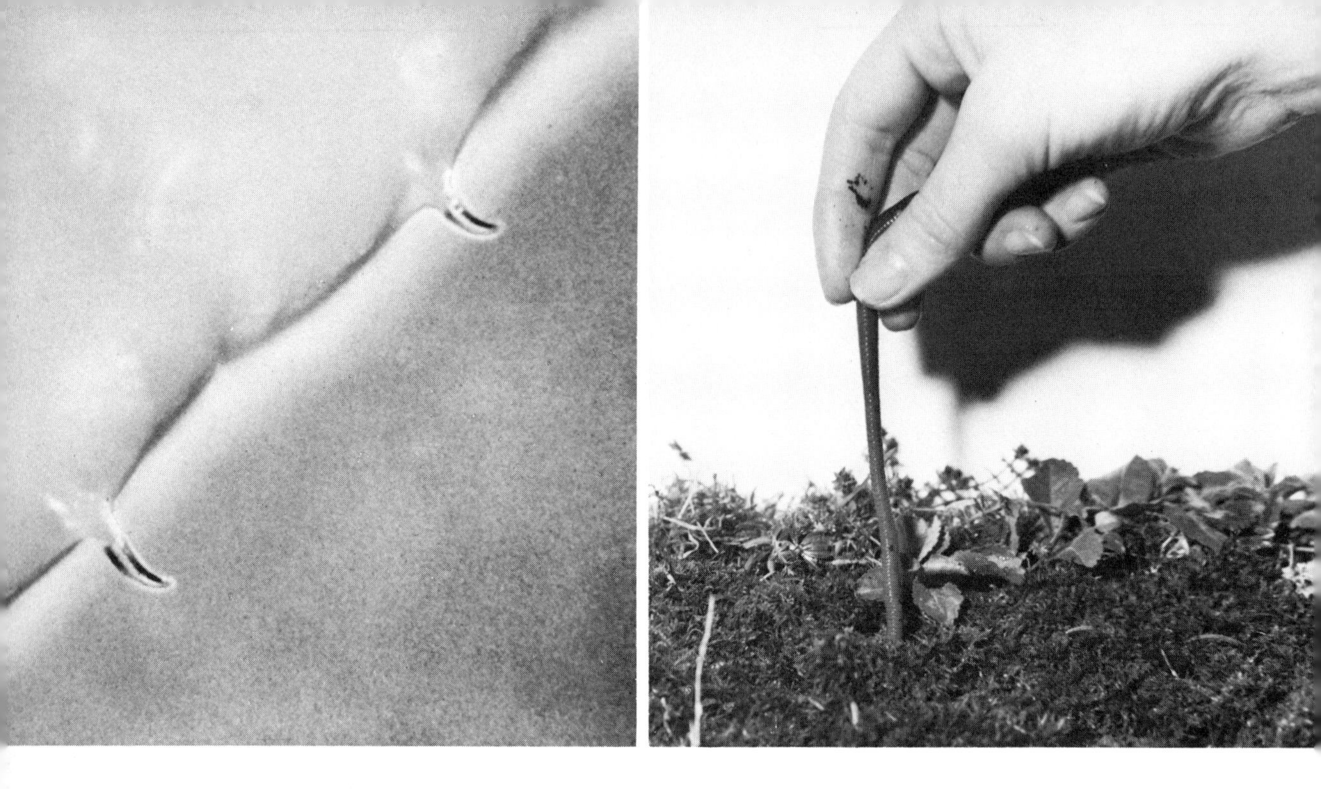

An earthworm also has four pairs of very tiny bristles on each of its body rings on both sides. The bristles are called setae and they help the worm to crawl or to anchor itself firmly in its burrow. Try to pull a worm out from its burrow and you will see how tightly it can hold on.

The earthworm is a terrestrial animal, which means it lives on land; yet it has not accomplished true land status, since it is restricted to an underground existence—a life of burrowing in damp soil.

When the atmosphere is cool at night and the evaporation of moisture is at a minimum, some earthworm species emerge and crawl about, but they never venture far from their burrows. Other species may spend most of their lives in burrows.

When the soil gets warm or during dry or hot spells, the earthworm goes deeper in the ground, where it can still find a little dampness. In the winter it burrows below the frost line.

Earthworms have special glands in their body to keep their skins moist. (Truly terrestrial animals have skins designed to keep the body from drying out and they also have lungs to breathe with.) Earthworms breathe through their skins, much like the way aquatic animals get their oxygen, and they therefore can live submerged in water quite a long time. If their skin dries out, or if they are exposed to strong light for long, they will die.

Sometimes you will find dead earthworms in a shallow puddle of water after a rainstorm. Rain probably filled the worms' burrows and forced them to the surface, where they got injured by excessive light or where the stagnant water did not hold enough oxygen for them to breathe. It can also be assumed that the worms in the puddle were already ill or weak.

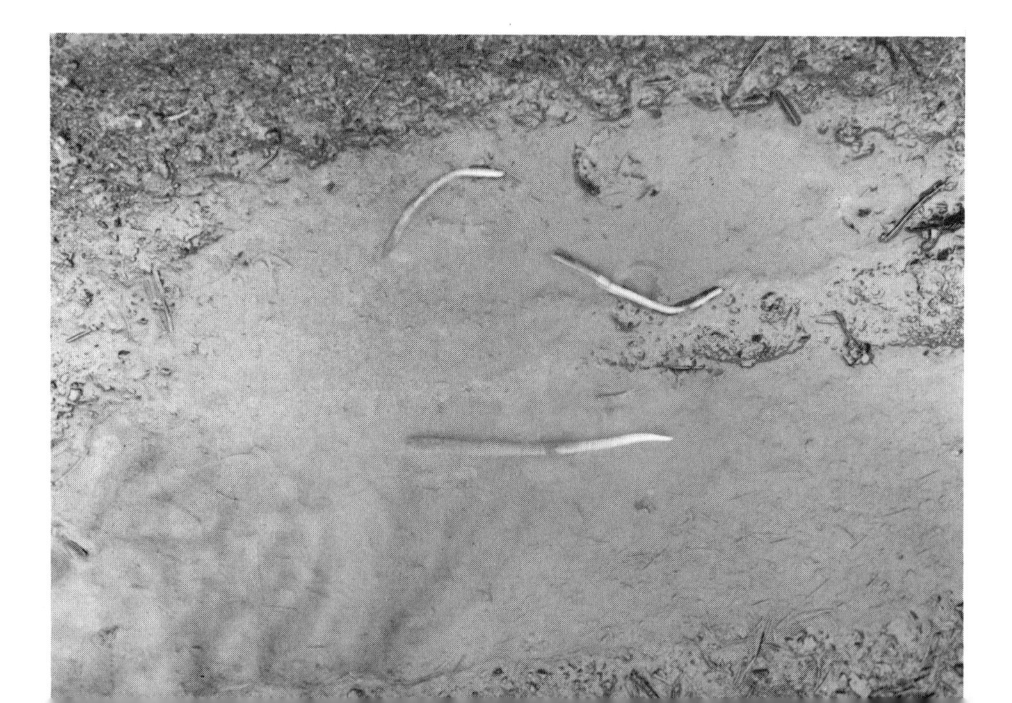

Since earthworms are nocturnal animals that come to the surface only after dark, they also do most of their feeding at night. They eat all kinds of organic matter, such as decaying vegetables, leaves, bits of roots, or grass cuttings. They also eat dead or decaying animal or insect matter. They pull this food or pieces of it into their burrows and swallow small bits together with some soil.

It is hard to believe, but a worm has taste cells and shows a definite preference for certain foods. You can find out which food a worm prefers by giving it little pieces of celery leaves, carrot tops, and cabbage leaves. Most worms will eat cabbage leaves if no other food is available but they will probably prefer the celery leaves if both are offered, and if they have a choice of all three food items, they seem to prefer the carrot tops. Try to feed them bits of fruit and different vegetable matter and see in what order they consume the food. Worms prefer food that is rotted. They also like oatmeal, cornmeal, chick starter food or laying mash, and even coffee grounds.

An earthworm has a mouth but no teeth; therefore, it cannot chew its food before swallowing it. Muscles push the soil and food the worm swallows through the esophagus to a chamber or sac called the crop. The crop stores food for a short time; then the food passes into the gizzard. The gizzard has strong muscles which grind the food, aided by tiny stones from the soil which the worm has swallowed with its food. (Chickens grind their food similarly.) The well-ground food now passes into the intestine where special glands secrete juices to digest it. Some of the digested material is absorbed into the bloodstream and distributed to the worm's body; the rest passes out

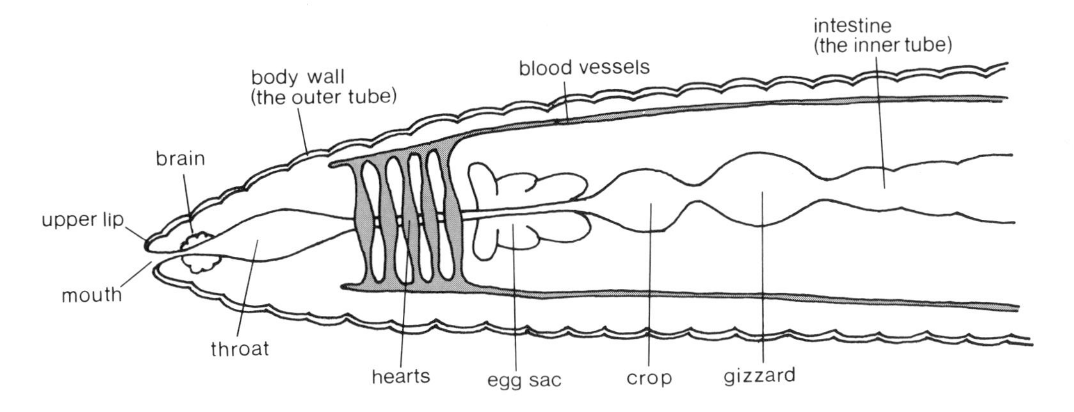

at the hind end of the worm as waste. Those wiggly-shaped little piles most of us have seen on lawns, in flower beds, or even on a dirt road or a driveway are worm manure, called castings. The castings make fertilizer which is rich in nitrate,

phosphate, and potash and is very highly prized. Some people grow worms only for the castings, which are sold to farmers, home gardeners, and plant nurseries. Worms can ruin a concrete walk by depositing their castings on it. Weeds will quickly grow there and cover the walk completely. Worms do not deposit all their waste on the surface; much remains inside the soil, in manure or compost piles, and under stones and in rock crevices.

It has been estimated that the 2 or 3 million earthworms living on one acre can produce up to 12 to 18 tons (about 11 to 16 metric tons) of castings in 45 days.

Worm castings keep wet soil loose and prevent it from crumbling and packing down so hard that plants might find it difficult to grow well. Worms and their castings also help the soil to absorb rainwater more readily; this tends to prevent excessive runoff of water. Water that runs off always carries good topsoil away and can eventually leave an area barren.

To see how worm castings help prevent soil from crumbling too fast, take two glasses of water and drop a worm casting into one and a little piece of plain sod of the same size into the other. The piece of sod will break up much faster than the casting.

Although earthworms help improve soil, they cannot change really bad soil; as a matter of fact, they will not even live in it. If you dig out a piece of sod 12 inches square by 7 inches deep and can find ten or more worms in it, it means that the worm population has a beneficial influence on the soil. If you find only two or three worms in the sod sample, it means the soil is not right for worms. Stocking it with new ones will be useless unless the soil is analyzed and improved first. Soil that has a good worm population is made even richer and more productive by the presence of the worms.

You can perform an experiment to demonstrate how beneficial earthworms are to growing things. Take two plants of the same species, age, and size and plant them in two separate, identical pots. Fill the pots with identical soil mixture. Add a few earthworms (two to four depending on the size of the pot) to one pot, none to the other. Put the pots side by side and water them with the same amount of water, always at the same time. After a month or two there should be a difference in the size of the plant containing the worms. It probably will be taller, bushier, and healthier-looking.

An earthworm has no eyes or ears, yet it can "hear" very well and "see" all it needs to see. It has light-sensitive cells, most of them in the front and tail ends of the animal. Animals that live in darkness often have no eyes but can find their way by other senses. African termites do an amazing amount of work without eyes; one species of cave fish in Mexico lost its vision generations ago; and animals like the mole have only minute eyes and very poor vision by our standards.

You can easily test how phototropic, sensitive to light, a worm is. Suspend two flashlights about 15 to 20 inches above a smooth-surfaced board or table. The lights should be about 2 feet apart. Cover one of the flashlights with green or red tissue paper or cellophane, leaving the other light uncovered. Place a few earthworms under or near the clear beam and watch them wiggle quickly toward the red or green area, away from the clear white light. The colors red and green are not brightly visible to worms and many other animals. Now try the reverse. Place the worms under the colored light. They probably will move in all directions or bunch together, but will avoid the

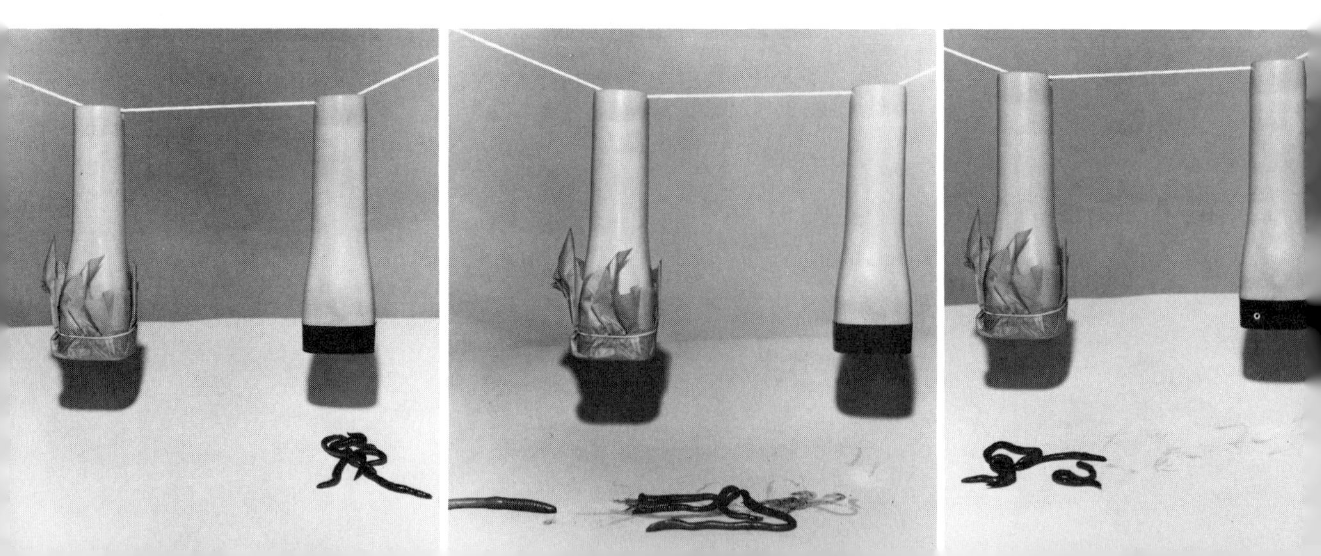

clear beam area. (If you see that your worms are getting dry during the experiment, moisten them with water.)

The absence of ears does not impair the earthworm's "hearing." It is extremely sensitive to sound vibrations. Some people claim that they can catch worms by pushing a stick or pipe into the ground and either hammering on it or moving it back and forth, the vibration causing the worms to leave their burrows.

Worms are very sensitive to touch, which you will notice if you poke one just a little.

No one knows how much pain, if any, a worm feels when it is put on a fishhook. It has no vocal cords and cannot cry out in protest, it can only wiggle. It has a well-developed nervous system, five hearts, and a very small brain.

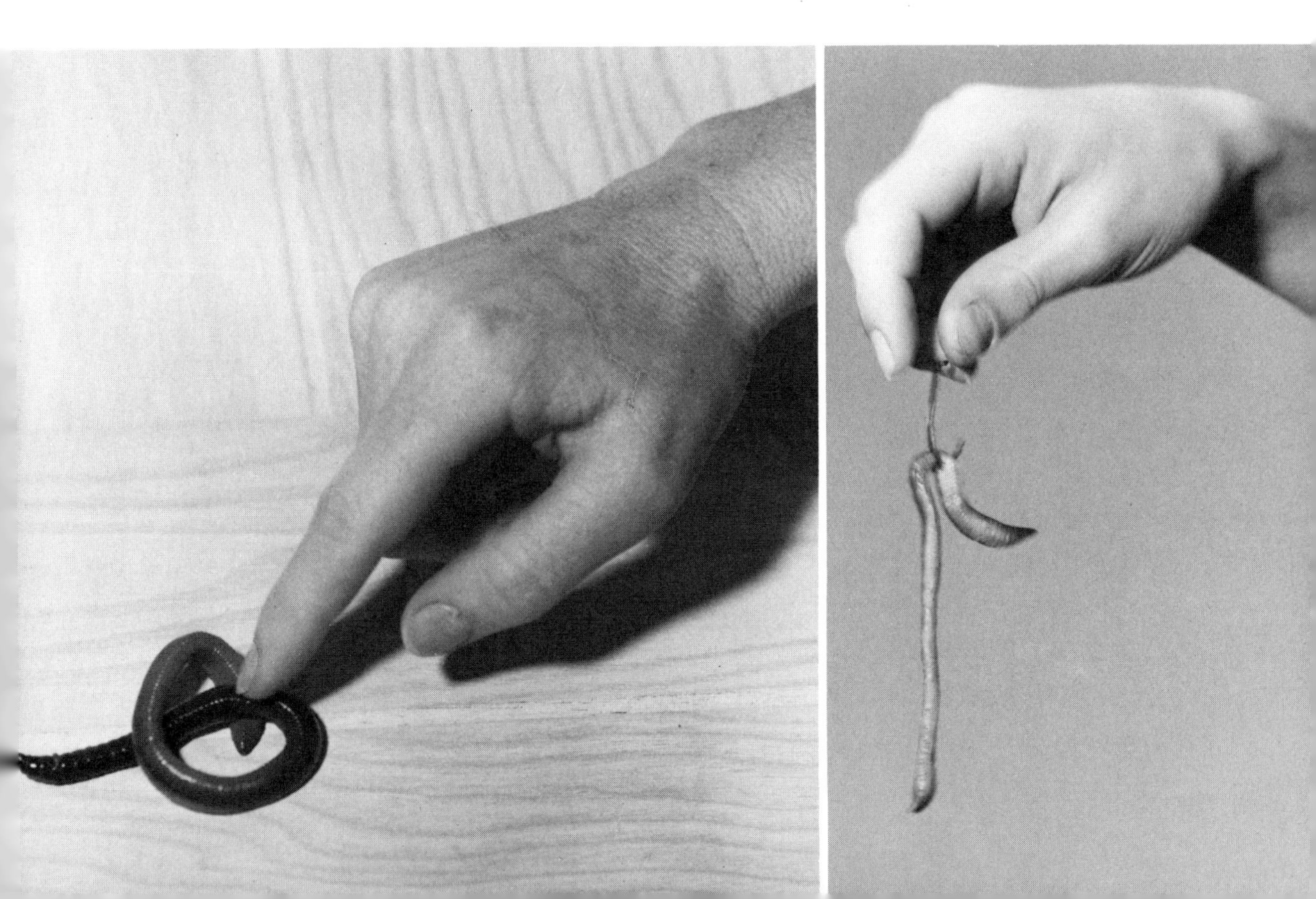

Although the worm's brain is tiny and primitive, it can be used to a certain extent. "Intelligent" is really not the right word to describe the actions of an earthworm, and yet even such a simple creature is capable of learning. To test how quickly your worm can learn, take a Y- or T-shaped test tube, or a test-tube connector that will form a Y or a T. (You can get it from a science lab, or a pet store that has tropical fish equipment.) To the lower end of the Y or T attach 8 to 10 inches (about 20 to 25 centimeters) of clear plastic tubing big enough for a worm to crawl through. Place a piece of cotton or cork in one upper arm of the Y or T to block the exit; set the other arm into a small pile of good moist soil. Wet your worm a little so that it can slide easier, and place it headfirst in the bottom part of the tubing. After several trials and errors, the worm will learn to select the right turn to escape into the soil. See how long it takes for a worm to learn this. Keep the worm moist during the tests.

In the spring, when the soil is warm and moist, the earthworm's sexual season begins; it lasts throughout the summer. Like snails, earthworms are hermaphrodite. That means each worm has both male and female organs in its body, but it usually needs another worm to deposit sperm in a sperm receptacle. Some worms may travel quite a distance to mate, others may select the worm living next door. If a worm mates with a worm in the next burrow, both worms may remain anchored with their tail ends in their own burrows, protruding only the

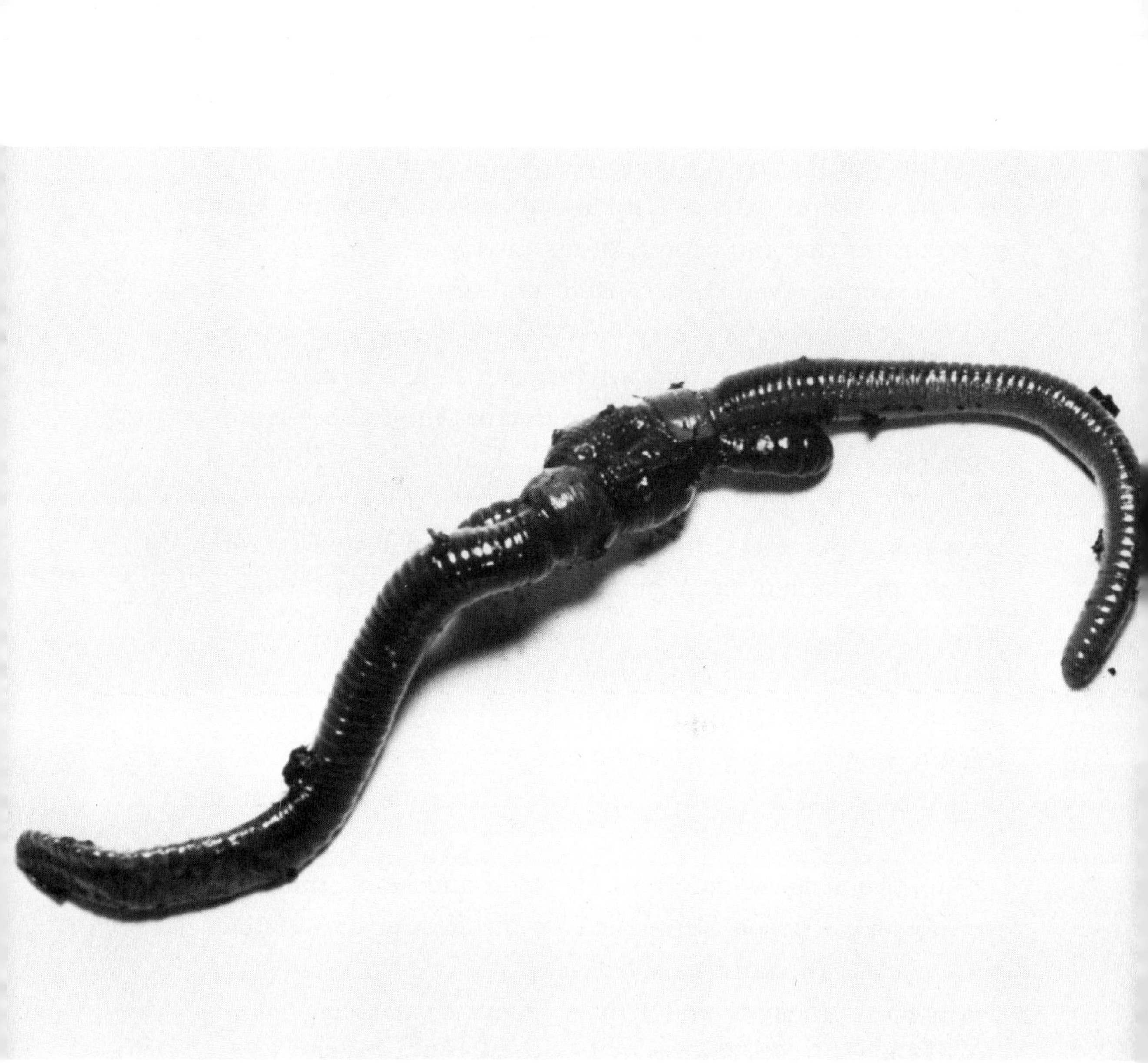

front end. Mating worms put their belly sides to each other but one faces one way, the other faces in the opposite direction. A tube of mucus holds the worm together.

A mature earthworm has what looks like a wide thick band around its front end near the head. This swollen ring is called the clitellum, and it contains a great number of glands which secrete a mucus ring shortly (a day or 2) after mating has taken place. The ring glides forward over the body of the worm, and as it passes over the reproductive organs it catches several tiny eggs and the sperm cells. The eggs are fertilized inside the mucus ring. The ring keeps moving forward until it slips over the front end of the worm and falls off. As it slips free, both ends become sealed tightly and form a little yellowish capsule. The capsule is sometimes called a cocoon. It is soft and rubbery and may contain from 1 to 20 eggs, but 4 or 5 is the average

number. The capsule lies in the soil for from 14 to 21 days, when the baby worms hatch.

Young earthworms are very thin, whitish, and about 1/2 to 1 inch long. You may never have noticed them in the soil when you dig for worms or do gardening, although they are probably there. If you want to see them, try burying a cob of a cooked ear of corn after you have eaten the kernels in soil, in a manure pile, or in a compost heap. After a few days, when you dig it up, there should be several baby earthworms feeding on it among some larger ones.

Earthworms mature to breeding age in about 90 days, but they will keep on growing for at least another four to six months. An earthworm is supposed to live as long as a dog, about 10 to 12 years, but most worms meet with an accident before they get very old.

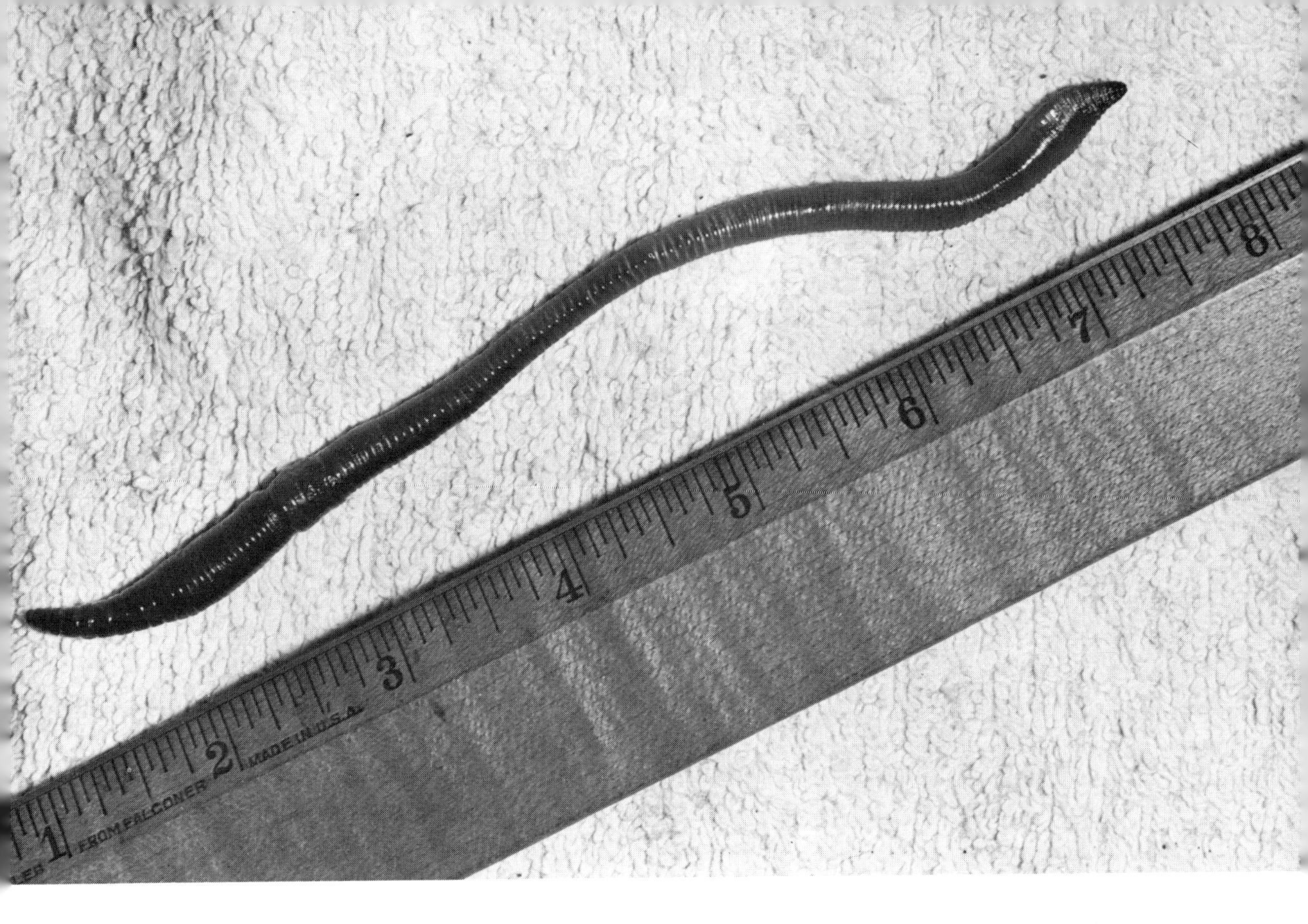

In old legends the word "worm" often referred to formidable dragons or serpents, but today it describes only small, wiggly, crawling creatures, which are often maligned or even feared. If a person is called a "worm" it is not meant as a compliment, even though the worm is such a big help to man. Charles Darwin, the great English naturalist who lived from 1809 to 1882, wrote a book about earthworms just one year before his death and in it he said: "It may be doubted if there are any other animals which have played such an important part in the history of the world as those lowly organized creatures."

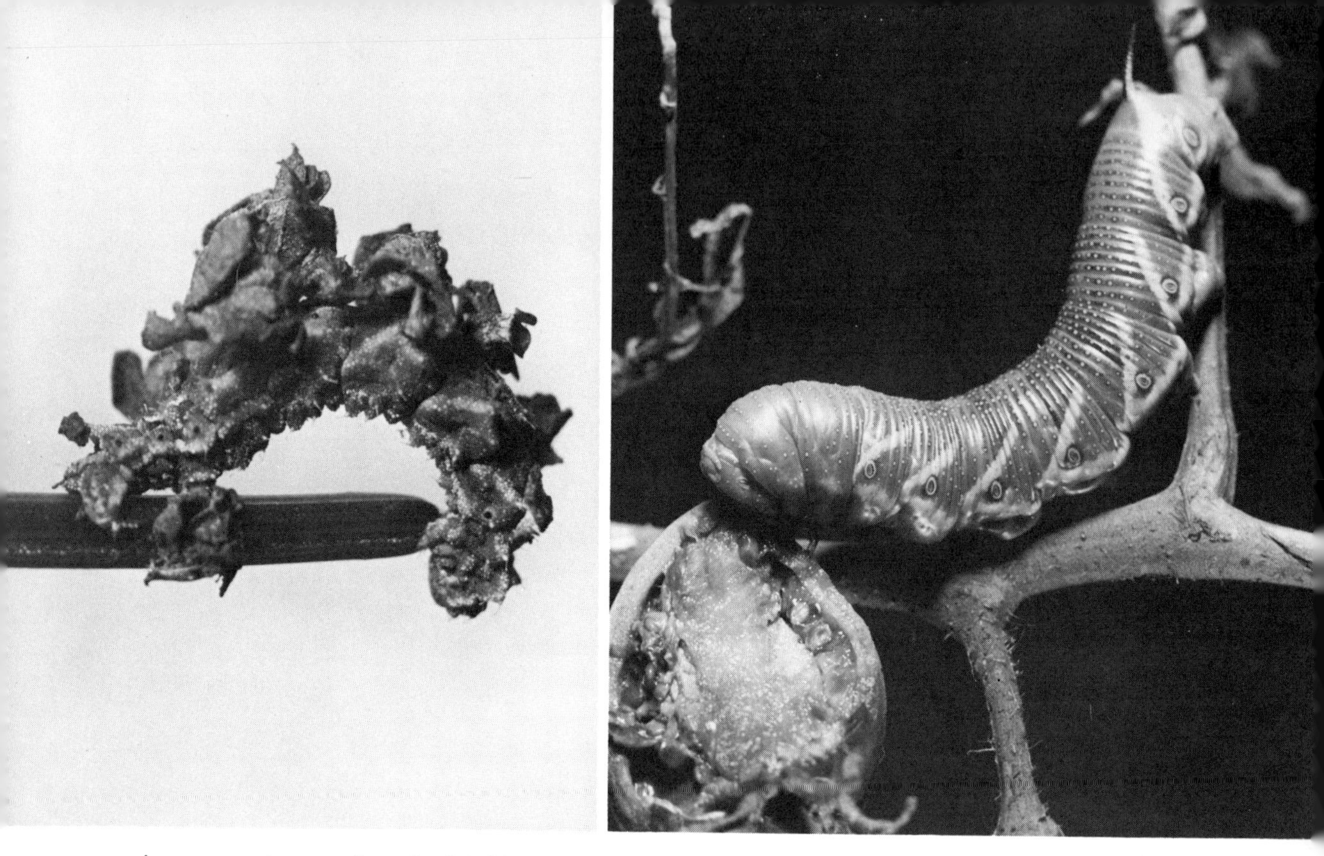

measuring worm (camouflaged with bits of leaves) *tomato worm*

Not every animal that is called a worm is a real worm. We wrongly call many insects, such as some caterpillars or beetle larvae, worms. The parsley worm, tomato worm, cutworm, mealworm, apple worm, inchworm, and many others are not worms at all. You can tell the difference by looking to see if the creature has legs. All but the true worms have legs, even if the legs are very tiny. Real worms have no legs at all.

All insects go through several stages of development. They lay eggs and the larvae hatch out; in some species they are called caterpillars. The caterpillar grows bigger and bigger and when it has reached maximum size it changes into a chrysalis. Sometimes the chrysalis has a protective cocoon around it; sometimes it is found in the soil or hangs on twigs or weeds. After having lived a while in this stage the adult insect emerges and it looks very different from the larval stage. It may be a beautiful butterfly, a moth, or a beetle. A real worm lays eggs too, but a fully developed, tiny replica of the adult worm emerges from the egg and it never changes into anything else. It just grows larger. If the animal is an insect it has three stages of development; if it is a worm it has only two stages.

promethea moth *earthworms hatching*

Earthworms have many enemies, but their greatest foe is man. It is not the worms collected for bait or fertilizer that constitute the danger to the worm population, but the worms which are destroyed on a very large scale by spraying poisonous insecticides on the land. Rain washes the sprays into the soil and they can kill every worm in it.

Natural enemies are numerous but they do not endanger the survival of the worm population. A mole eats its own weight in earthworms every 24 hours and hoards worms by the

thousands in its underground burrow. Birds, lizards, frogs, toads, skunks, rats, snakes, gophers, and even ants and centipedes eat earthworms. Fish will eat some water worms, but earthworms are not their natural food—yet we all know how much fish seem to like earthworms. In Australia, one of the oddest animals in the world, the platypus, lives almost entirely on earthworms, and if a zoo wants to exhibit a platypus, a lot of earthworms have to be grown specially for it. The electric eel from South America is also fed many earthworms in captivity.

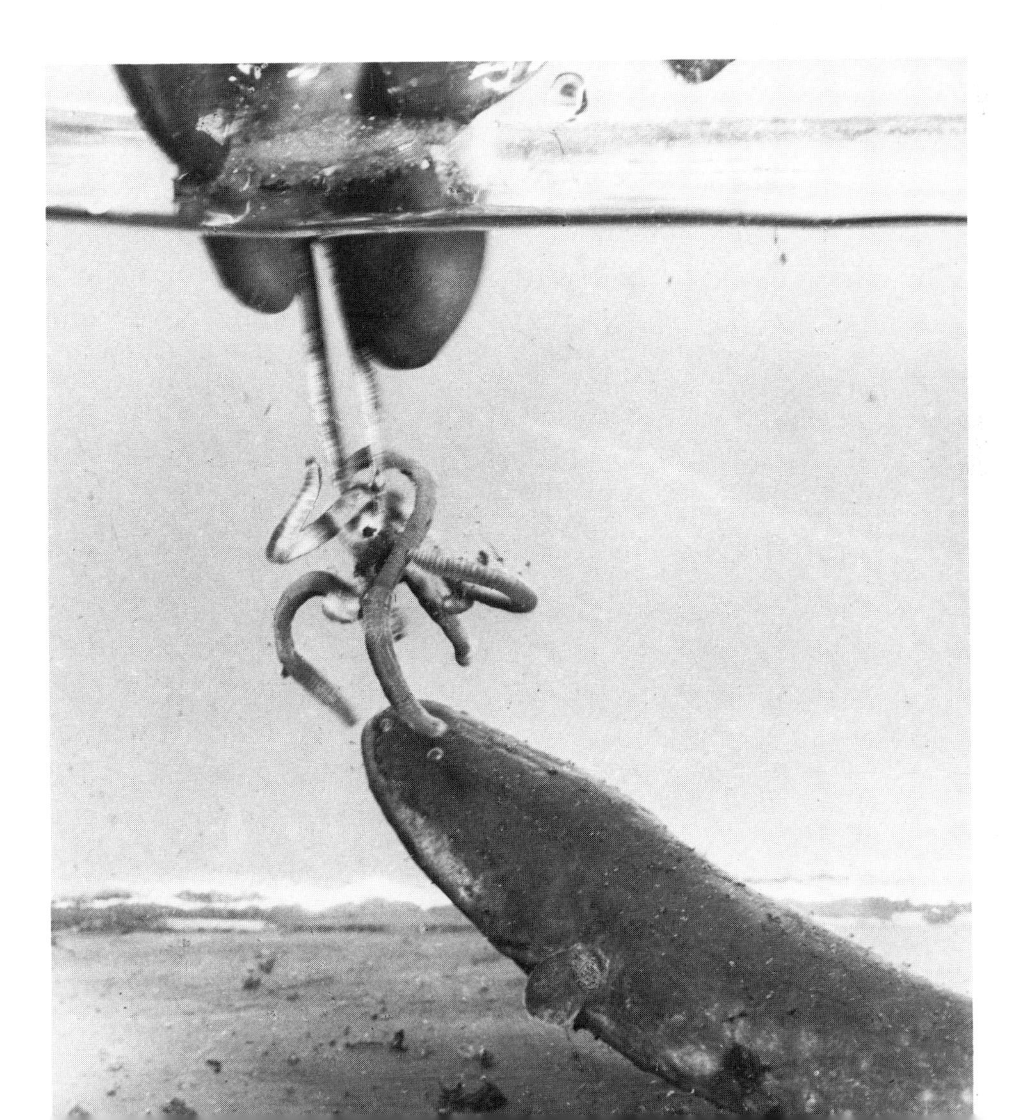

Earthworms are very rich in protein and when dried can be added to some pet and cattle foods. They are eaten by humans in some parts of the world, mostly in underdeveloped countries. Some laboratories are experimenting to find ways to extract the amino acids in worms for human consumption. Amino acids are the building blocks of proteins, and the blocks can be com-

MAUREEN RUFE/THE POCONO RECORD

bined in any number of ways, just as the letters in our alphabet can be combined into any number of words. A fad to promote earthworms as food has recently started in this country. Cooked worms are supposed to taste crunchy and chewy. In California 2000 recipes were submitted in a "worm recipe" contest. There were entries for worm cake, worm cookies, worm casseroles, and worm pancakes. Now many local worm growers sponsor contests for the best worm recipes. The little children's song—

Nobody loves me
Everybody hates me,
Goin' out the garden
To eat worms.
Big, fat juicy ones
Long, slim slimy ones
See how they wiggle and squirm!
Bite the head off,
Suck the juice out,
Throw away the skin.
Nobody hates me!
Everybody loves me!
See? I won again!

—could become obsolete if worms became a delicacy. As of now they are not an accepted item at mealtime.

If a person's head were cut off, it would mean instant death, just as it would for all other mammals, birds, fish, and reptiles—although some reptiles can grow a new tail if they lose the old one through an accident. Some forms of life, however, can grow new heads, new tails, or new arms. The ability to replace the lost parts of the body is called regeneration. Lower invertebrates such as protozoa, hydras, sponges, and planaria can often regrow a complete new animal provided some vital nuclear cells are still present. Higher invertebrates, such as the starfish and the earthworm, can regenerate only certain parts. A starfish can grow a new arm if it loses one in an accident and an earthworm can grow a new head or a new tail.

If the worm's head with just 9 or 10 segments is cut off, a new head will form. If more than a quarter of the worm's front end is removed, the worm will die. If a quarter of the tail section is lost, a new tail and some new segments will grow. The cut-off piece of the tail will not form a new complete worm, nor will the cut-off head. Thus you can't get two worms from one this way. If a worm is cut in half or cut lengthwise, it will die. It takes an earthworm from two to four weeks to grow a new head or tail. If the worm loses its head again shortly after a new one has formed, it may be too weak to regenerate a third one since it has not been able to eat for a long time. A tail regenerates more easily, but if it is removed too often the worm will either die or just heal over the wound and live with a stump of a tail.

A distant lower relative of the earthworm, the planaria or flatworm, lives in water and is a master of the art of regeneration. It can not only replace a lost head or tail, but can grow many new heads or many new tails. It is often used in the classroom or laboratory to demonstrate this remarkable ability.

The planaria is a flat, small worm with a triangular head and very primitive eyes which can see only light and dark. It has a nervous system and a minute brain. You can usually catch planaria by placing a piece of raw meat on a string in an unpolluted pond, stream, or spring. After about an hour there should be many planaria feeding on the bait.

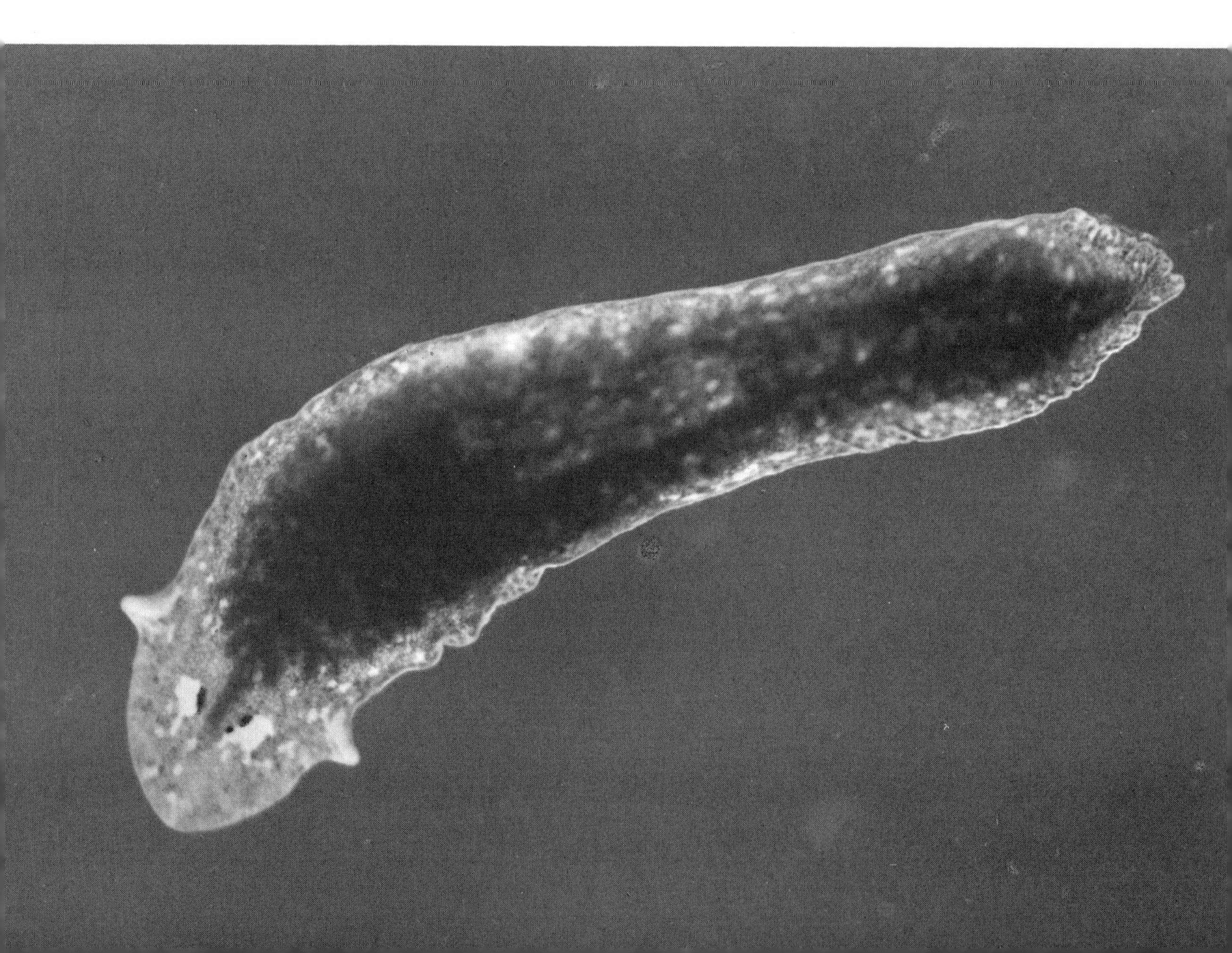

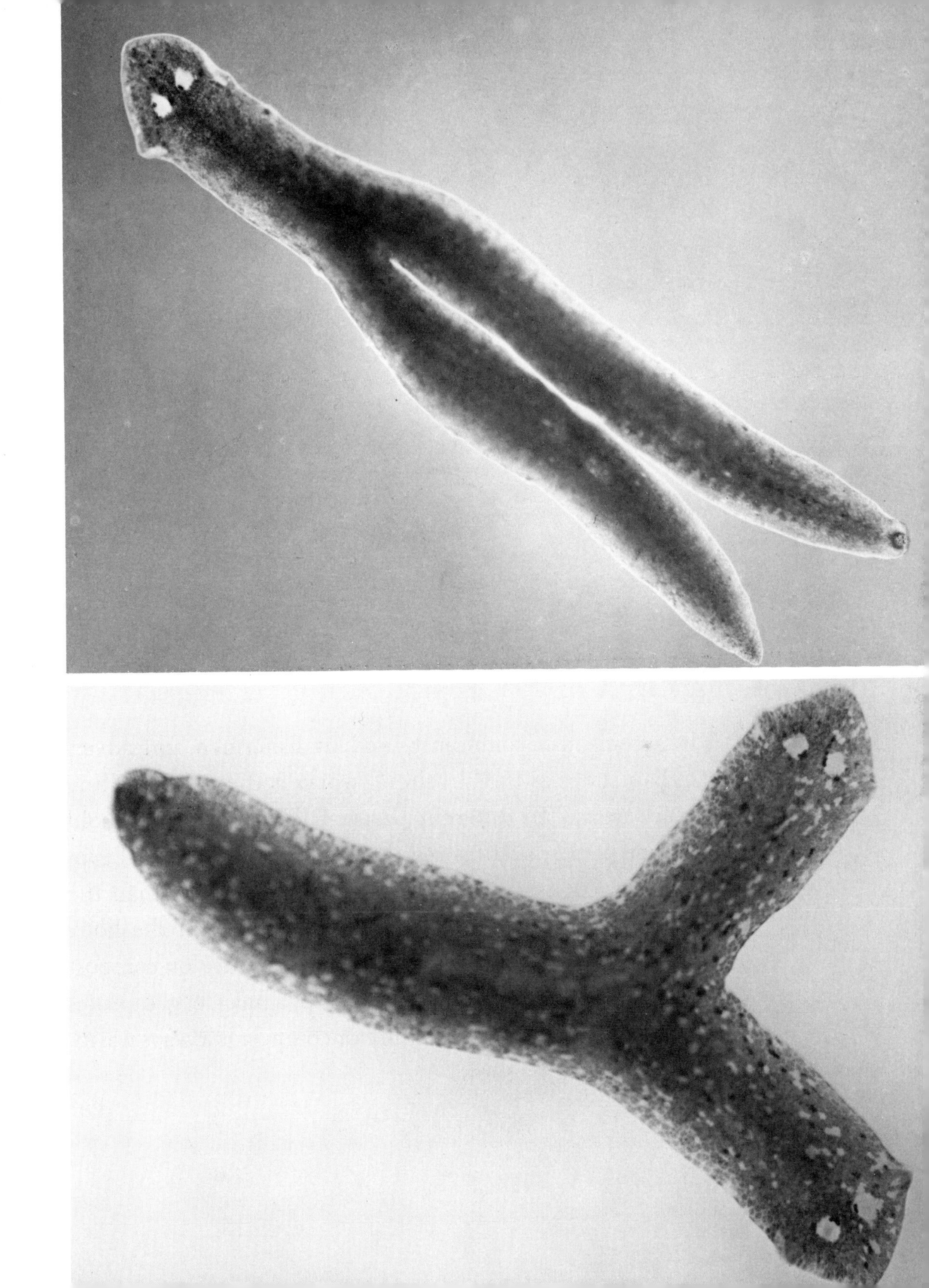

The worm most commonly used by fishermen and advertised in bait shops as the "liveliest" worm is the red brandling or manure worm. In different parts of the country it is sold under other names, such as red wiggler, California red, Egyptian red, and English red. This red worm is smaller than the ordinary garden worm and it lives much longer on the hook and under water. It can be found in any manure or compost pile. It grows quickly to maturity and produces egg capsules all year round because moist manure or compost is always warm, even when covered by snow.

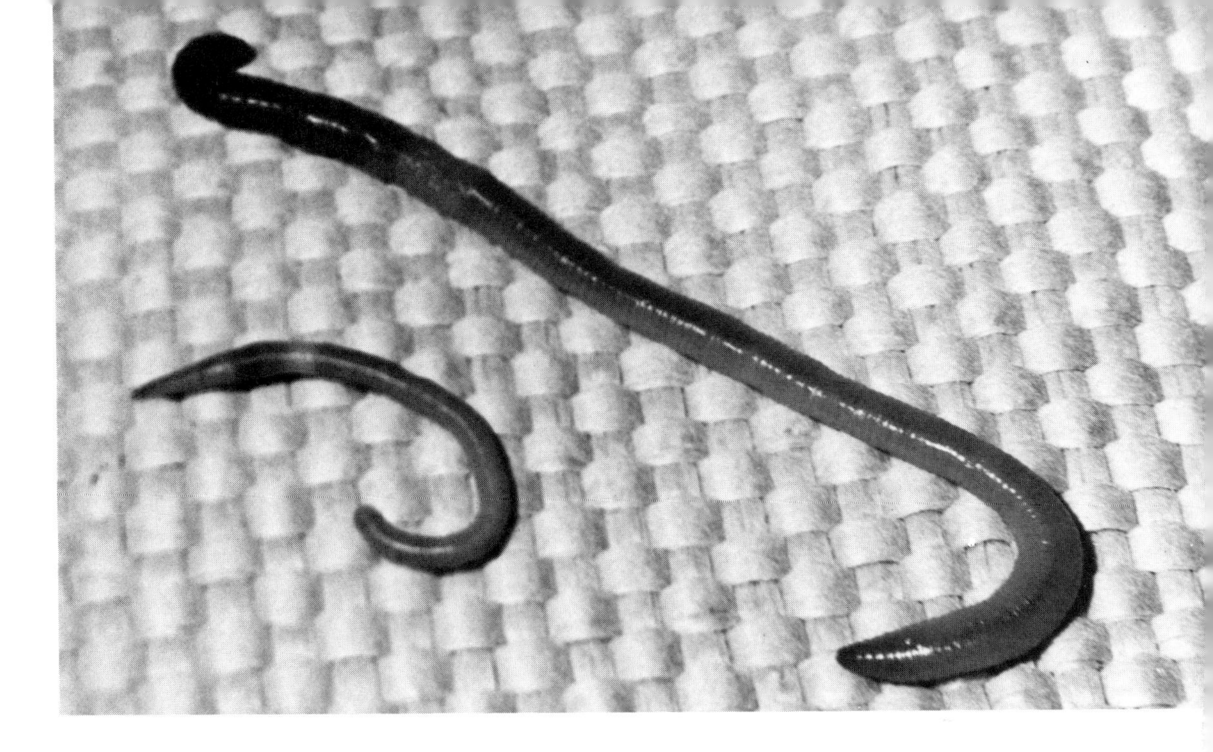

Common garden worms or angleworms are the ones people often dig for themselves before they go fishing. These are the worms that hungry robins pull out of the lawn. We all know the saying "The early bird catches the worm." The worm caught by the robin was tardy and did not retreat all the way into its burrow in time.

Nightwalkers, or night crawlers, are our largest native worms and also great favorites for fishing bait. They originated in Africa. Because of their nocturnal habit of crawling about at night, you can catch them by the beam of a flashlight and pick them up. On a damp night or after a rainstorm there are usually many worms wandering and feeding outside of their burrows. Walk quietly and slowly, since the worms can feel the vibration and will quickly slide back into their burrows.

People raise earthworms for bait, for fertilizer, or for breeding stock to sell to newly starting worm growers. New uses for the earthworm are constantly explored. Lumber companies are experimenting to see if earthworms can speed up the growth of new trees in soil where old trees have been cut. Since worms consume all sorts of garbage such as vegetable and human wastes, paper and wood products, experiments are being conducted to use worms in refuse disposal sites. The worms dispose of almost all the garbage quickly and as a bonus they leave valuable fertilizer and many new worms behind. But the garbage needs to be mulched first, and a way has to be found so that worms can work through the winter months in colder regions.

Commercial earthworm farms are big business. They are located mostly in the warmer parts of the country where the worms can live outdoors year round. An earthworm farm may sell about 750,000 worms a week, rising to as high as 700,000 per day during the peak fishing season. Boxes containing about 5000 worms are shipped to retailers, who sell them by the piece or put them up about 10 or 12 worms at a time in cartons. In some areas you can buy worms in worm vending machines. A little container holding about 15 to 20 worms can be bought for $1.

Worm raising is not only profitable for the large commercial growers, but also for young people or retired persons who like to have a little extra income. A worm business can begin very small and expand later as needed or desired.

BAIT
LIVE BAIT
24
HOUR
SELF
SERVICE
SPRING WATERS
WORM FARM

Many booklets are available to tell you how to set up a large or small worm farm. This can be done in the home, in a garage, in a garden plot, or in large outdoor pits. In the home wooden boxes are used, outdoors the pits can be made of wood or concrete. Your local library probably has most of the literature you need to set up a good-sized worm operation. Or you can write to the U.S. Department of Agriculture in Washington, D.C., for their free booklets on worm farming.

But as a small hobby or for observation in your nature study, you can raise some earthworms in an old fish tank or a plastic or wooden box. A box 18 by 15 inches long and about 7 inches high is very good for a starter worm culture. If you want to expand, just stack several boxes on top of each other.

Drill a few small drainage holes in the bottom of the box and line it with window screening, and on top of the screening place five or six pages of newspaper. Mix equal amounts of sand and rich soil and fill the box two-thirds full with the mixture. On top place a layer of well-rotted leaves mixed with some peat moss. Moisten it all with water and put the worms on top. Do not use more than 20–25 adult worms as a start. The worms will dig into the soil within minutes. You can cover the worm culture with damp burlap or more moist newspaper to prevent it from drying out too quickly. Your worms will live and breed in this mixture for a long time, but they will do better and get fatter if you feed them. Sprinkle a thin layer of poultry mash or cornmeal on top of the soil and scratch it in lightly. Green leafy

vegetables, grass cuttings, potato peelings, a small amount of coffee grounds, and almost any other small amount of fruit and vegetable matter are eaten by worms. Do not feed them too much; start with just a little, and if it is all eaten within a few days you can increase the amount slightly. Never let the soil dry out, but don't keep it soaking wet. Use a plant mister and dampen the soil daily.

If you want to raise red manure worms substitute a mixture of manure or compost mixed with an equal amount of peat moss for the sand and soil mixture. This mixture too should be kept moist but never soggy.

Some stores sell styrofoam boxes made for fishermen who want to raise a few worms. They come with a package of ready-mix worm bedding and a smaller package of worm food, plus instructions.

After a month or two it is time to check for eggs. Dump out all the bedding and remove the large worms, which you can put into a new culture. The old bedding with the cocoons (or capsules) and the tiny white baby worms can be put back in the original box where they now will have room to grow and develop. Another method of removing the large worms from a box is to stop feeding the worms for a few weeks and then place some food in one corner of the box. You can probably scoop up most of the large worms in the evening when they surface to feed. Naturally there will be some large worms left in the culture, but it will contain mostly the young worms and the cocoons. Worm farmers have special machines that sort out worms by size.

The worms you need to start your worm culture can be bought or you can collect them yourself. If you buy them, get "pit-run" or "bed-run" stock. These are unsorted worms, containing eggs, baby worms, and adult worms of all sizes. They are the most economical worms to get. If you dig your own worms, use a garden fork rather than a spade or shovel, which may injure too many worms.

When you have raised more worms than you need, return the excess to the garden or field so they can continue to keep our soil healthy and productive.

Earthworms are all around us, in our backyards, woods, and fields, and it is interesting to observe them and to record their busy, secretive lives.

INDEX